The Death of Earth as We Know It

William (Bill) C. McElroy

Copyright February 2020
ISBN 9798627169224

Preface:

We live on a little blue marble that is floating around in space amid trillions of other chunks of floating rock and ice. Our planet by chance or design is situated just far enough from our sun to keep us from freezing to death or being burnt to a crisp. We lucked out in that certain elements and gasses decided to embrace us and provide us with all the necessary items we needed to survive and reproduce. Unfortunately we as a thinking mammal have found ways, many, many ways to destroy the gift we inherited. This short booklet was written in an effort to affect your conscience by making you aware of the consequences of 'cause and effect' that may be our downfall and extinction, if we do not wake up!

Author:

I have been an advocate for science and technology since I could first read. Living on a farm in the *Catskill Mountains* of New York in the 1940-1960 eras I learned to hunt, to farm, to fish, to explore, and to accept and appreciate nature every day. My parents provided me with a 21-volume set of Compton's Encyclopedia, which I read cover to cover several times as television was not readily available during those years.

I joined the USAF at age 18 and was trained as an *ECM* {*Electronic Counter Measures*} technician assigned to the Security Service {Pre NSA}. After the USAFSS I spent a year as a registered scientist on a *Columbia University Oceanographic vessel*, and then a year at *Mercury Nevada* testing atomic bombs. I am now in my late 70s, and have visited many nations, states, museums, parks, and environmentally sensitive locations. The text you are about to read is an accumulation of observations and experience. Enjoy

Table of Contents:

Chapter # 01 – Weather Related

Weather vs. Climate seems to be a path of contention and confusion among those that are trying to describe what is happening to our planet. *Weather* is local; it can be isolated to an area as small as a few miles or as wide as an entire nation. *Climate* is worldwide and is a measure of the overall temperature and weather trends. Our climate is heating up, and this heating is causing changes in our weather patterns both nations wide and in many instances locally.

Global Warming:

Hoax of Real, the debate continues as the scientists measure the earth's temperature and plot the rise in temperature over the decades. Yes, we have had major global warming caused by natural causes in the past. The earth had periods of time when volcanoes were prevalent and the earth was very warm. We had times when the earth was nearly frozen from pole to pole, only to eventually warm and become tropical swamp.

Ice Age:

Many of our scientists predict a new ice age is in the making and their theory is based on the following. The warming of the earth allows more water to be evaporated into the atmosphere. This increase in water vapor when cooled produces larger volumes of snow and ice. More snow and ice results in the sun's light being reflected back into the atmosphere and into outer space, instead of heating the ground. This is a self-feeding action that produces more snow and ice until the warming of the oceans no longer is possible, and thus the oceans freeze and we have an ice age.

Dust Bowl:

Many of our citizens do not remember the dust bowl and how it caused massive migration. The Midwestern areas of the United States, mostly around the Texas Panhandle and Oklahoma were tree-covered forest ripe for settlement. The settlers were farmers and to farm they needed rich fertile soil. To obtain this soil they leveled the forest and then planted crops that took months to germinate. Meanwhile the sun was drying the soil and the rains were few and far between. In May of 1934 the winds {*Aeolian processes* (wind erosion)} came up and lifted the dried earth into massive dust storms that made life unbearable and the land no longer suitable for growing. This caused a massive migration of people from the area back to the safety of the east coast and to the dreams of a better life west of the Rockies. The period that is considered the *Dust Bowl Years* is from 1934 to 1940. {1934, 1936, and 1939–1940}.

Expanding Water due to Heat:

When water is heated it tends to expand as it also does when frozen. It is good that water does expand when frozen as it thus floats, and forms icebergs that stay on top of the oceans. If water did not expand it would sink and thus may forever remain as layers of ice under the oceans, rivers, and lakes, which is not a good situation.

When water heats it does expand, but it has been proven that water at 0 to 4 degrees *Centigrade* tends to be stable. Below that it freezes and expands, above that it heats and expands. Thus, if our oceans and large bodies of water heat above 4 degrees C each will tend to expand. If you expand millions of cubic miles of water, i.e. an ocean, it will cause the water to rise upward and thus, cover beaches, islands, and other low lying areas like coastal cities.

The *rising oceans* and lake waters due to heating or cooling can thus block the flow of water from the higher elevations to the lower elevations, and this backup can cause flooding that can cause damage to man-made structures. The most recent example of this was 2017 when *hurricane Irma* along with high tides, and higher than normal sea levels caused the rivers to backup and flood major portions of Charleston, South Carolina.

In *Florida the sea level* rise has caused sea water to flood the streets and contaminate the freshwater supplies. The state and city of Miami are currently spending $400,000,000.00 in an attempt to save the city and its precious water supplies.

World wide the estimated cost to defend against sea level rise destruction to man-made items is approaching $14 trillion dollars; and this figure is going up every day.

Fires and Fuel:

Water causes fires; yes or no? The fact is that the more water we have in the atmosphere the more snow and rain we will have and thus the more plant life that will be produced. Plant life is considered a good thing as it cleanses the air, produces oxygen that we as humans require, and absorbs nitrogen and other gasses and chemicals that may be somewhat harmful to humans. The problem is that with the increase in plant life comes the increase of dead and dried up plant life in the fall, winter, and early spring months. This dead and dried up plant life is a perfect fuel for massive fires as recently seen in California and some of the western states. {2018-2019}.

Permafrost:

The melting of the permafrost is an indication of serious things to come. It means that the earth is warming, and the overall climate and weather are changing. This leads to major changes in the production of food and living conditions for millions.

What is Permafrost?

When the ground stays frozen at or below 0 C or 32 degrees F, for a period of two years or longer it is said to be 'permafrost'. Frozen ground can be walked on, driven on, built on, and in areas of the Arctic and Antarctic there are towns and other structures built on the permafrost.

Melting Permafrost:

When the permafrost melts it no longer can support structures like homes, stores, factories, etc. These buildings will slowly sink into the muck as the ice that held it all together melts. This is already happening in parts of the Arctic where native villages are pulling up stakes and moving from miles to hundreds of miles as they resettle on firmer land.

Permafrost that is between the ocean waters and land are becoming too soft to travel upon, thus making it difficult for natives that depend on the ocean waters for seafood, furs, and other marine items.

Rising Oceans:

The oceans worldwide are rising due to two situations. The first situation is that the landmass under the oceans is rising in some areas and thus, pushing up the water levels. The second instance is the melting of ice across the globe. The ice caps, the glaciers, the icebergs, and the permafrost are melting at an alarming rate and this has resulted in trillions of gallons of water being added to the sea levels.

This *sea rise* is harmful in several ways starting with the flooding over of low-lying islands, coastal areas, etc. The second major problem is that adding freshwater to the oceans is changing the chemical composition of the oceans, and thus the composition of the marine life that we require for a food source. The third problem is that in places like Florida, the sea rise is affecting the *groundwater* and turning fresh drinking water into arid water no longer suitable for domestic use.

Melting Ice Caps and Glaciers:

If it snows and the snow never melts due to being in a place that never heats, then the next snow will cover the first snow. As snow after snow builds up it tends to compress and melt just enough to turn to ice. Over the centuries this process produces ice that can be from feet to nearly a mile thick.

As this process progresses over the years there are dust, bugs, animals, and other materials that get trapped in the layers; this includes viruses and bacteria. While trapped these items stay inert and present no harm to human life, but when exposed due to melting, may.

We also have *scientists* working at tunneling and drilling into the depths of the ice to study the warming and cooling of earth, which can be done by looking at the thickness of each layer of ice {snow}, and this process may someday release something into society that may just be detrimental to our way of life, or may actually help us solve a way of killing off some virus or bacteria. These scientific studies can go either way.

The melting of a *glacier* or ice cap is self-generating in that when the surface melts the water has to go somewhere, and if that water finds a crack it will fill that crack, thus warming it. This then expands the crack and makes it deeper into the ice, and when it hits solid ground it flows under the ice. This under the ice flow acts like a lubricant that allows the ice to break loose from its ice to ground anchorage and thus start moving downward toward the sea. Many of our glaciers and ice caps have now moved miles and some have completely disappeared into the sea in the form of icebergs.

The amount of ice on earth is vast and if it all melts the volume of water will raise our oceans by many to hundreds of feet, thus flooding low lying lands for dozens to hundreds of miles inland.

And here is where the balance of nature takes over. *Global warming* will cause more water evaporation, which in turn causes more snow to fall, which in turn when very cold will form ice in the form of glaciers and ice caps. But, if too much warming occurs, then the air will not become cold enough to snow and thus, the new ice will not be formed and old ice will melt and the seas will rise. Additionally, all the 'stuff' that has been trapped in the ice for thousands of years will be released into the oceans, the results of which are unknown.

Additionally, as the ice melts the weather will change and this will affect growing seasons, marine life, ocean flows, and coastal temperatures, etc. The prospects of what can and probably will happen are enormous and in most instances detrimental to human and animal life.

Chapter # 02 – Pollution Related

Pollution can be in the air, the water, on land, in the soils, and at the bottom of the oceans. Pollution causes many problems to mankind such as loss of food, illnesses, and death. Some pollution causes problems immediately; much of it causes problems over periods of time that can be from weeks to years.

Chemical Pollution:

We use tens of millions of tons and gallons of *insecticide*, *pesticide*, *fertilizers*, and other chemicals that have a positive use in that each is designed to solve a problem. We also are polluting our earth and its land and waters while doing so; thus the question is which is worse, the problem to be solved or the solution?

Environmentalists will usually seek alternative methods of controlling insects, ridding vermin, or making things grow faster and better. *Commercial farmers* tend to go the other way and use the chemicals that make things easier for themselves and increase profits.

Recent farming practices are to plant vertically in enclosed buildings that are temperature and lighting controlled; use recycled water spiked with fertilizers, and keep the building free of insects and human pollution.

Petroleum:

Oil is found throughout the world, in some places only a drop or two that is not worth the expense of capturing, in other areas millions of barrels, thus making the drilling, piping, and transporting worthwhile. In the United States we started with oil from Pennsylvania, moved to Oklahoma, then Texas, then California, into the Gulf of Mexico, onto Alaska, and now in the Arctic. We also have started and fought wars in North Africa, the Middle Eastern nations of Iraq and Kuwait, and we have spent countless trillions searching for new sources of oil over the last hundred years.

Petroleum is used for making jet fuels, gasoline, diesel fuel, plastics, some cosmetics, lubricating oils, and much more and therefore it may take decades for new materials to replace it. But, it will have to be replaced, as you can see we have moved further and further from the original Pennsylvania site in our quest for oil. Eventually we will find ourselves in a situation where there just is no more or where the little left must be reserved for the very rich and our militaries.

Crude oil has to be refined into the various forms of petroleum products and gasses and to do this it must be transported from the well to the refineries, sometimes over thousands of miles. Pipes expand and contract with the change of temperature, hot to cold and cold to hot and this expansion causes cracks and broken pipes. These breaks dump the oil onto our land and into our rivers where it pollutes, kills birds, fish, animals, and makes the land unsuitable for growing food.

In the *Gulf of Mexico* we have 'dead zones' that are caused by industrial and farming runoffs of chemicals, and by the decades of oil leaking from the thousands of capped off wells. In short, we have and are killing off our marine life and our seafood production.

Gulf of Mexico, Chesapeake, and the Delaware Bay:

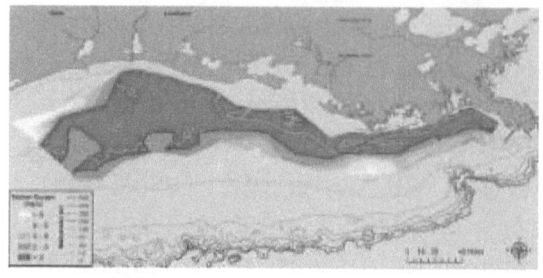

The below picture is from NOAA and shows the extent of the *Dead Zone* {Zone with low or no Oxygen in the water due to chemical pollution from lawn, street, and farm runoff}

Dead Zones:

Per NOAA

> "*This large dead zone size shows that nutrient pollution, primarily from agriculture and developed land runoff in the Mississippi River watershed is continuing to affect the nation's coastal resources and habitats in the Gulf.*
>
> *These nutrients stimulate massive algal growth that eventually decomposes, which uses up the oxygen needed to support life in the Gulf. This loss of oxygen can cause the loss of fish habitat or force them to move to other areas to survive, decreased reproductive capabilities in fish species and a reduction in the average size of shrimp caught.*"
>
> "*Scientists have determined this year's Gulf of Mexico "dead zone," an area of low oxygen that can kill fish and marine life, is 8,776 square miles, an area about the size of New Jersey. It is the largest measured since dead zone mapping began there in 1985.*"

REF: noaa.gov   Gulf of Mexico Dead Zone is Largest ever Measured

Per NOAA

> "*June 13, 2016 Scientists expect that this year's mid-summer Chesapeake Bay hypoxic low-oxygen zone or "dead zone" – an area of low to no oxygen that can kill fish and aquatic life –*

will be approximately 1.58 cubic miles, about the volume of 2.3 million Olympic-size swimming pools. This is close to the long-term average as measured since 1950."

Quotes from the Experts:

"Monitoring how nutrient levels may be changing in streams, rivers, and groundwater and how the estuary responds to these changes is vital in assessing our overall progress in improving the health of the Bay. Local, state and regional partners rely on the basic data to inform adaptive management strategies in Bay watersheds."

REF: said *Don Cline*, USGS associate director for water

"There has been a recent trend toward less hypoxia later in the summer that may signal an emerging response to actual reductions in nutrient pollution, but it's no reason to be complacent — we have a long way to go to finish the job."

REF: *Donald Boesch, Ph.D.*, president of the University of Maryland Center for Environmental Science

There are also *Dead Zones* in Green Bay, Wisconsin, Delaware, the Arabian Sea, Caribbean coast of Panama, Bay of Bengal, and other areas around the world. Many nations are now seeing the negative results and are trying to do something about each. In short, our *Coral Reefs*, and Fishing areas are dying and we humans seem to be the cause.

Suffocated spots

{Picture is from the World Resource Institute –Each red dot indicates a Dead Zone. }

"Hypoxia (low oxygen) isn't even mentioned in several of the most important academic reviews of threats to coral reefs and is rarely discussed at

scientific meetings, even worse, many coral-reef monitoring efforts do not include measurement of oxygen levels, making it nearly impossible to identify low oxygen as the cause of mass coral mortality after the fact."

REF: *Andrew Altieri*, staff scientists at the Smithsonian Tropical Research Institute

Natural Gas and Fracking:

For over a hundred years we have thought of natural gas as a waste product of oil exploration and many of the oil wells burned the natural gas as it exited from the ground. In the 2000s someone finally decided that natural gas was a commodity that could be harnessed and sold as a replacement for coal and fuel oil. After all, it burns cleaner and is reasonably inexpensive.

There are a few *problems with natural gas*, starting with the fact that it does burn, and if allowed to collect in one spot can explode. Natural gas explosions have destroyed several buildings in California, Philadelphia, and in the Boston suburbs in the past few years. Gas piping in many areas is now decades old and leaking and will have to be replaced; a costly process.

Additionally, the new method of obtaining natural gas is 'Fracking' where a hole is drilled in the earth, water is injected into the hole under high pressure, and the earth fractures allowing the trapped gasses to congregate and then be collected. This process is known to cause *earthquakes* in areas like Oklahoma where earthquakes did not exist before the Fracking operations.

Thus, Fracking operations have allowed trapped gasses that were deep in the earth to rise to the surface. To do this the gas has to travel through the water table and thus, the water we use for irrigations, washing, and drinking is getting polluted {mixed} with the *flammable gas*. People have been known to set their tap water on fire due to this.

My parents lived in upstate New York for most of their lives and had absolutely some of the best drinking water one could obtain. A neighbor drilled through the water table and into a Sulphur deposit, and thus every well in the area became a stink hole smelling of *rotten eggs*. It cost thousands to install a system to counter this, and the water never again tasted pure. This is what is happening to citizens that live in the areas of the nation where Fracking is being done, and it is causing health and economic problems.

Coal:

Coal for generations provided the fuel for heating homes and businesses, and also was an inexpensive fuel for the generation of electricity. Unfortunately, burning coal pollutes the air and the coal ash waste pollutes the waterways and the land.

There are currently many political and social attempts to eliminate coal fired power plants due to the fact that people are getting sick and dying from the pollution generated. The coal mining industry itself is dangerous and destructive, and there are miners getting sick or dying in mines worldwide. The coal industry has had over 100-years to check itself and come up with ways to prevent the harm done to people and our planet by itself; they have in short 'thumbed their nose' at our governments, health care providers, workers, and citizens and have thus brought upon themselves the activist movements to end coal production.

Chapter # 03 – Disease and DNA Related

As I am writing this, the *Coronavirus* is overtaking the earth and killing people by the tens of thousands. This is a *pandemic* that can wipe out millions, and can cause countless emotional, financial, and other harm to our populations.

Disease:

Viruses and bacterial diseases are with us every day. Most people develop *antibodies* that defend against these diseases, but some people are just too old, or unhealthy, or too young to fight off the onslaught of these tiny creatures that live among us.

We have over the decades managed to build up immunities to many diseases; to create *vaccinations* {Shots} for the prevention of many very harmful diseases, and we have learned how to protect ourselves from the spread of diseases. What we have not done is to insulate ourselves against all diseases, nor have we eliminated 100% all of the killer diseases.

A pandemic like the *Spanish Flu* of the 1918 era, or the current 2020 Coronavirus have created havoc upon our civilization and our economic welfare. The person to person contact spread of the disease is fast as we now live in a society that is very mobile, and saturated with people in every theater, store, ballpark, aircraft, subway, sporting event, and political rally.

There are both good and bad with a pandemic; the bad is that people die, people get sick, panic results, and lives are overturned. The good is that people die, people get sick, panic results, and lives are overturned. This current outbreak is a wake up call for our medical professionals, our governmental officials, and our private business communities. It shows how strong and how weak we are against an invisible foe.

It creates situations that we have not thought of in depth, like should people attend weddings, funerals, events, etc., where people to people contact can result in the spread of the disease and possible death to the attendees. It creates a purpose that brings families together, and it creates a need for better family and social planning.

We have during the 2020 pandemic found that we need to change some of our *medical procedures* and laws; we need to figure a way to compensate the businesses that have contributed to the fight for health, and those that are going bankrupt due to lack of paying clients. We need to figure a way to feed our school children while our schools are closed, and we need a way to provide childcare for these children when the parent must go to work, i.e. firefighters, military, national guard, police, medical workers, government employees, and such.

We need to help our *tourist industries* as people stay home and away from trains, rental vehicles, RV purchases, motels and hotels, resorts, amusement parks, restaurants, national and state parks, ski areas, golf courses, and sporting events.

We need to help our *students* that do not have access to high-speed *Internet* so that he or she can complete his or her education at home. We need to figure a way to deliver food and medications safely to the poor and the elderly at their homes or in rest facilities.

Most importantly we need a central worldwide organization that can coordinate medical care, instructions, testing kits, vaccines, food, and other items that are needed for the health and safety of the public at large. The WHO {World Health Organization}, and CDC {Center for Disease Control} are two of the organizations that need to be better respected, and that need better communication practices with the emphasis on informing the general public on how to conduct their daily lives.

The 2020 pandemic may be helping to save our planet; it is cutting the population; it is grounding our polluting machines; it is showing us that we need less government spending on war and military might and more on people and their health. It is showing us that we are woefully unprepared for a pandemic, and it is showing us how interconnected our lives are with others, even those we despise or hate.

Travel Contagion:

We board the aircraft for a 30-minute flight from Burbank to San Jose. We are with 80 or 90 other souls and we expect the flight to be uneventful. Frankly, for the most part this route is like a bus ride, safe and secure and very routine for many. But, today something else has boarded with us, a super small, microscopic sized bug that can reproduce itself every few seconds. This bug, virus, is sitting in 29B and is now spreading throughout the aircraft via small airborne droplets from the host not covering his or her mouth while coughing. By the time the aircraft lands in San Jose all passengers plus the 5-person crew have the virus, and in a few days shall be very sick to the point of perhaps even dying.

Since no one was actually showing signs of being sick while on the aircraft the aircraft is not disinfected between flights and after a short stop at San Jose is loaded with another 80 passengers and is heading to Denver. Meanwhile the first group of passengers moved off down the hallways of the airport onto the baggage carousel area and then into the waiting taxis, Ubers, and family and friend's vehicles. It has only been an hour and already hundreds of people have been contaminated with the virus, albeit unknowingly. These people then over the next few days go shopping, to events, home to families, to work, and to dental and other appointments. By day two over a thousand people have been exposed, and at the end of thirty days over a million. The 1918 Spanish Flu killed 50,000,000 by destroying their lungs and thus allowing more virus and bacteria to attack the immune systems of the young, old, and everyone in-between.

Permafrost Germs:

As the atmosphere heats up and the permafrost melts there is a good possibility that some bacterial element that was frozen eons ago is released into our atmosphere and takes us all out.

Oceanography:

I worked on an oceanographic vessel for nearly a year back in the mid 1960s and one of the things we were doing was taking *cores*. A core is a 20-foot length of dirt and mud obtained by dropping a heavy pipe into the bottom of the ocean. As the pipe penetrates the mud it fills. The mud is then removed from the pipe and analyzed for content that may stretch back for thousands of years. We took absolutely no preventive measures to restrict possible biological material from contaminating the ship or the laboratories and storerooms back on land.

In addition to the cores we took water samples from water that was miles deep and may not have been near the surface for a million years. We also dragged wire nets along the bottom and scoped up rocks and minerals that lay on the sand or muddy bottom. Each of these items may have contained items or diseases that were millions of years old.

Germs from Outer Space:

We since the late 1950s and early 1960s have been visiting space; *Mars and the Moon* are two such heavenly bodies that we have been to and from which we returned. Our *NASA{National Aeronautics and Space Administration}* people do a reasonably good job in trying to keep our germs from traveling into space, and also keeping possible space germs from being brought back to earth. But, there are now the Chinese, Russians, and many other nations exploring space, and eventually one of these will be bringing back something from space that we really do not want here. Perhaps we should be taking the warnings that Hollywood has been providing us since the days of *Captain Video*, the *Twilight Zone*, and *Buck Rogers*.

Germ Warfare:

Does this exist? The chances are great that it does and governments worldwide are working on 'specialized' germs that can be used to win future wars by targeting specific groups of citizens. For example, using gene technology it may be possible to produce a germ weapon that only kills or incapacities one race and a specific age group in that race.

To counter this, our scientists and military are working to produce new virus species so that we can produce a counter-weapon species in the event that our enemies create a virus weapon species. Yep, the circle of life, or death continues.

DNA Change:

DNA or *Deoxyribonucleic Acid* is part of all life and it is the roadmap of molecules that create us with size, color, sex, and all the attributes that make us whom we are. In recent years our scientists have discovered the 'keys' to our DNA and how to manipulate DNA so that we can create new plants, new animals, and new humans. This manipulation has also allowed some scientists to create means of cures for some birth and accident defects.

Gene Therapy:

A part of our DNA is the 'Gene' that determines the color of our hair, the texture of our skin, etc., and these genes sometimes go bad and create disease and defects that we wish would go away. Scientists are now experimenting with replacing 'bad' genes with 'good' genes, thus curing all sorts of illnesses and extending life.

The problem with gene therapy is that a mistake can actually replace a bad gene with a worse gene that can duplicate and spread, thus making the person sicker or even killing him or her.

Mutations:

When we treat a disease we use various methods like genes, chemicals, antivirus injections, and other techniques that are designed to treat and kill off the disease. Disease like bacteria and viruses are living things, and we are the host or its home. If someone tries to remove you from your home you do what? You fight back, you object, and you do everything you can to stay put. Diseases are very much like you, each wants to live and that means not being forced out of the host, you!

The survival instinct takes over and the disease, be it a bacteria or a virus fights, and if necessary changes its shape, habits, needs, etc. This therefore creates a new species and the process is called 'Mutation'. When this happens we humans are forced to find another 'cure' and it may work or it may trigger another round of mutations that are even more difficult to kill. This is currently happening, as each time we come up with a flu vaccine the next season there is another variety of flu that requires another more powerful vaccine. Eventually, one of us has to win, the disease or the human, and it is not necessarily the human.

Sterile Males or Females:

Did you know that drinking water comes from pee {urine}? There are 6.5 billion people on our earth and many times that many animals and we all urinate several times a day. *Urine* is water, and that water is collected and recycled via city or town owned purification plants or via seeping through layers of earth until it enters the *water table*. A water table is a layer of water trapped deep within the earth and of which we store and then tap via drilling a well for our use.

We take medications, all sorts of medications for all sorts of illnesses and other personal reasons. Women may take *estrogen* to control hot flashes, and men *testosterone* for muscle buildup and sexual prowess. These *hormones* {chemicals} are used by the body and then discarded via excretions of feces and urine;

the hormones are then introduced into our water supply, and frankly not to many purification plants filter each out. Thus we are drinking water that may have concentrations of these chemicals, and each is changing our bodies to the point that a female may become more like a male and a male may become more like a female.

Hormones have been found in bottles manufactured from plastics, and in water contaminated by *Fracking*.

Food Additives:

Ever read the label on a package or can of food? Horrifying at best to see all the strange names of chemicals that have been added to the items inside. These are for color, appeal, longevity, taste, and all sorts of other reasons and are accepted by the *FDA* {*Food and Drug Administration*} as perfectly edible substances. Many people have switched to 'Organics' in an effort to no longer be fed these assortments of chemicals.

The thing is that these additives usually do have a reason for being added, in limited quantities are reasonably harmless, and in some instances actually helpful. For example, *Wonder Bread* was produced during WWII as a means for the average citizen to get their daily allowance of vitamins, which otherwise many would not have received due to the rationing of food at the time.

One problem with our food is that much of it, organic or not, is being grown in fields that for decades and maybe centuries have been fertilized with chemicals or with manure and these items are absorbed into the food being grown every year. In short our bodies are being slowly poisoned each time we eat.

Chemicals to Kill Insects:

To add to the poison in our food, our farmers have for decades added *insecticides* to the fields in order to kill off bugs that could ruin the crops. These chemicals are designed to 'kill', yes bugs, but over time just about everything that consumes each, including the animals we eat, and the plants grown in this chemical soup.

Runoff:

Each time it rains or snows and melts there are fields of chemicals that are washed into the streams, ponds, ditches, and puddles that surround the fields. This chemical enriched water then flows to the rivers and eventually to the oceans and bays. This in turn creates a 'soup' of chemicals that feed algae, and starve the water of oxygen that is needed for the growth of coral and other marine life, including the fish and shellfish that we eat. See *Dead Zones* in this manual for more information.

Chapter # 04 – Resource Related

Resources like water, food, minerals, chemicals, etc., are needed for our version of civilization to survive. People and animals need food, housing, clothing, and emotional and spiritual support. As the population grows, as transportation allows distance between relatives and friends, as communications allows even more distance, people are becoming more isolated, and animals more bold.

Food Depletion:

If you walk into a grocery store in the U.S. you are amazed at the variety and quantity of food that are available; but in some nations just a half-cup of rice and a sip of water is an amazing thing. Life requires both subsistence and water to continue, and many portions of our earth are barren of both.

Population Overload:

As the populations of both animals and humans increase there is and will be more and more battles over food. Where an animal may in the past only had one *predator* enemy, today it may have several hungry predators seeking a meal.

The human population is not just competing for food and clothing, but also for land and housing. The *Real Estate market* displays this competition in that a home that sold for $18,000 in the 1960's may now be offered and sold for $180,000 or even more. Some high population areas of our nation have homes that should be selling for thousands being sold for millions.

This need for land has taken away valuable produce and grain farmlands, dairy farms, and low cost housing areas.

Food prices have gone from approximately $6.50 per grocery bag in the 1960s to $30 per bag or more in the year 2020.

Gasoline has gone from $0.17 in 1960 to as much as $5.00. It is today around $2.50 per gallon, but if you work out the cost per mile driven, that is a major increase per mile. Example: One gallon at $0.17 in 1960 took you about 17 miles, or $0.01 per mile. Today the $2.50 takes you about 25 miles, or $0.10 per mile a ten times increase.

Growing Area and Seasons Change:

As the earth heats up due to population increases the land that once produced 'xyz' crops can no longer be productive due to the temperature changes that are not suitable for the particular crops.

This is leading to more areas that were once off limits having forest cut down, land plowed up, and more chances of disease being released, and animals being displaced.

The *Endangered Species Act* (*ESA*) in 2020 listed over 1,890 species that are going extinct, this is added to the hundreds of thousands that have already gone extinct in the last decades.

Why does this matter, well it is because there is an interrelationship between animals and plants, and many to most of the plants we rely on for food require some interaction with birds and other animals. Animals tend to control dangerous insects; feast on animals that feed off of crops; and pollinate crops so that each can bear fruit. Without predator animals and other species, many of our edible food crops would vanish from the face of the earth.

Mutations:

Some of our food producers, some of our scientists, and probably some of our enemies have been or currently are experimenting with *'Cloning'*, DNA manipulations, and other forms of mutation in an effort to produce new foods, new weapons, or new species. In nature it was thought that mutations took hundreds to thousands of years, but this has been proven incorrect. One instance is the nectar feeding birds that have chosen to start eating seeds. The long beak that is used to extract nectar from flowers has in only a few years shorted and thickened to the beak of a seedeater that needs to crack open a seed husk with its beak.

Many creatures in nature are changing, mutating, to other forms as food supplies are changing due to climate and weather changes. Many are changing due to interrelationship crossbreeding, and others due to needing to cohabitate with man. For example, squirrels, mice, rats, and lizards have learned to live in the same homes as humans. Dogs have over the centuries changed from wild animals that may have eaten humans to domesticated animals that are now 'Man's best friend".

What we need to worry about is the 'frog' that mutates into a dinosaur that is 30-foot long and enjoys eating humans. Can it happen, yep in all probability it can.

Water Depletion:

Water, Water, everywhere but nary a drop to drink; every sailor understands this as on the open oceans one is surrounded by water that if drunk can result in death due to the salt content. Much of the earth's water due to hundreds of reasons is now poisonous to humans and many animals.

Water Ownership:

We are seeing the results of water ownership in that bottling water plants are draining local community wells dry, thus leaving the communities forced to 'purchase' the bottle water from the companies that are basically stealing a natural resource, i.e. the community's water.

Salt Water Contamination:

Florida and other towns near the oceans are seeing that rising sea levels are allowing salty sea water into the freshwater tables, thus contaminating the community water supplies. Florida is spending tens of millions of taxpayer dollars attempting to not just stop the salt water breaching, but also to purify the water that is already contaminated.

Water Industrial Contamination:

Companies like *Ford Motor Company*, some battery manufacturers, and dozens of mining and chemical companies have over the decades thrown waste materials into valleys, streams, rivers, and lakes. Most that are still in existence have been caught, fined, and made to clean up the mess they created, but many went out of business or are bankrupt and the destruction left behind has now been designated as '*Superfund*' sites, that are costing taxpayers billions for the cleanups.

Some areas like the '*Love Canal*' area near Buffalo New York are so polluted that no one is permitted on the land, and the housing project that was built on the land had to be torn down, the owners displaced.

Oil drilling and production, Fracking, coal plants, and other fossil type energy production have polluted many streams, lakes, rivers, and bays. The *Gulf of Mexico* is a '*Dead Zone*' due to pollution, i.e. not capable of sustaining any marine life.

Low Rain Production:

Why is a *desert* dry? The main reason is lack of water in the form of rain or snow. An additional reason is the location that prevents clouds of water vapor from changing into rain or snow. Another reason is that there is a lack of vegetation that produces local weather that generates rain or snow. If you cut down all the trees, and only have grasses or dirt, then you have created a situation where moisture is not stored and therefore, not evaporated into the atmosphere, i.e. a desert.

Cutting Down Trees:

Cutting down massive amounts of timber releases CO_2 {Carbon Dioxide} into the atmosphere and that release traps H_2O {Water} in the atmosphere that traps heat near the surface that creates *Global Warming*. It also reduces the amount of tree generated Ox {Oxygen} into the air, i.e. oxygen we humans need for sustaining life.

Cutting down massive amounts of timber causes the earth to warm, as there is no shade cover, and thus no cover to keep rainwater from evaporating. This in turn dries out the land even more, and this dry land then blows away in storms or washes away during a storm, leaving no place for stored water. A forest stores millions of gallons of water, keeps the earth cool so that rain can form, and provides roots that hold the land in place. Without a forest none of these conditions will happen and we will end up with desert, and a major lack of rain and snow for plants, animals, and human consumption and use.

Lead Pipes:

Pb or lead has been around for generations and it was used in paints, and for making pipes. Lead is soft and malleable and easily formed into pipes, sheets, rounds, and other shapes. The plumbing industry for decades used lead pipe, lead traps, and lead fittings because it was thought to be safe, easy to work with, and inexpensive. Well two out of three is not bad; except for the fact that it turned out that lead was not safe and actually harmful to humans as *Flint Michigan* discovered.

From Pure-Earth.com

> *"Health Effects of Lead in Drinking Water. This damage commonly results in behavior and learning problems (such as hyperactivity), memory and concentration problems, high blood pressure, hearing problems, headaches, slowed growth, reproductive problems in men and women, digestive problems, muscle and joint pain." www.pure-earth.com/lead.html*

How does lead affect the body?

> *"Key facts. Lead also causes long-term harm in adults, including increased risk of high blood pressure and kidney damage. Exposure of pregnant women to high levels of lead can cause miscarriage, stillbirth, premature birth and low birth weight." www.who.int/en/news-room/fact-sheets/detail/lead-poisoning-and-health*

Chemical Attack on Water Sources:

Most city and township water supplies are monitored for chemicals that can be deadly to animals and humans and if any such chemical is detected the alarms sound and hopefully the water supply is turned off. Much of the water in the east is from lakes high in the mountains and it is piped down to the towns and cities below.

Smaller towns have wells or use stream water that is filtered and purified. The problem is that there are literally hundreds of chemicals that can be harmful and that are not being detected or monitored, and to a terrorist this fact becomes an opportunity for mischief that can sicken or kill. The only thing that tends to prevent this from happening is the vast amount of dangerous chemicals that are needed may be difficult to obtain and kept from detection before being added to the water supply.

Biological Attack on Water Sources:

A biological attack is easier than a chemical attack in that a biological agent can multiply itself while in the water supply, and therefore a gallon of dangerous biological material could possibly become hundreds of gallons as the water it is in flows to the population zones. Again, big city water supplies tend to monitor for biologics and other contamination; whereas smaller towns and neighborhood water supplies may not have the facilities for doing so.

Water Depletion:

Another danger is that the water supply dries up and therefore, a town, village, or city may be left high and dry without any water. Many places around the world have limited supplies of freshwater and therefore, the water has to be brought in as bottled water. *Bottled water* is available at most U.S. grocery stores, but do you know from where it is obtained? Some U.S. towns have run dry of drinking water due to the bottled water company drilling wells and pumping 100% of the water out, thus leaving none for the town.

The *Coca-Cola factory* in Mexico, bottle water plants owned by *Aquafina* and *Dasani*, and the *Nestlé Water Bottling Plant* in Sacramento are all causing major problems due to their draining the water tables with their wells and pumping. This demand for certain water {Taste, smell, clarity, and mineral content} has devastated towns and cost the locals dearly as the lack of water for washing, drinking, and bathing is depleted.

Chapter # 05 – Politics Related

The United States of America has some 40 political parties, of which three, the Independents, the Republicans, and the Democrats are the prevailing. The problem is that we really do

NOT have any parties, but rather social economic groups that do have different lifestyles and needs.

Republicans vs. Democrats:

For example, the majority of Democrats and maybe half the Independents live in big cities and socialize as mixed groups that have learned to live among each other, well almost.

The Republicans and perhaps half of the Independents tend to live in rural areas where farming, hunting, fishing, and other outdoor activities are prevalent. They tend to be spread out and socialize in churches, town halls, local bars and coffee shops.

The income levels between the groups can vary widely, although there many multimillionaire farmers, most of those living in the rural areas are not considered wealthy.

Guns and Gun Owners:

The rural communities tend to treat guns as recreational items and items used for controlling varmints {groundhogs, squirrels, snakes, wolves, coyotes, rats, etc.}, whereas the city gun owners tend to treat guns as weapons of power, weapons of protection, and collectible items. Hence, the great gun debates of the 2000s that do not seem to have an answer or suitable conclusion as tens of thousands are hospitalized or shot dead each year.

As this debate roars on, the groups are continuously arming themselves with more lethal weapons, and millions of rounds of ammunition as they brace for the impending civil war.

Gun and Ammunition Sales:

The *Coronavirus* scare has people worried about feeding their families as food is disappearing off of store shelves due to 'hording' by those that feel they are more important than anyone else. At the same time, gun and ammunition sales are skyrocketing, due to first the season of the year and secondly for protection against neighbors that may want the food that is in your pantry. Again, it is the *ME, ME, ME generation* that will be the downfall of our civilization.

Rich vs. Poor and Medical Help:

The poor have more access to health care than the rich in some instances. No one by many state and federal laws may be turned down for emergency medical help, regardless of status or income. This sounds great, but in practice does not live out to its claims. For example; *dental* is not covered by the VA {*Veterans Administration*}, *Medicare*, or *Medicaid*, even though poor dental work can lead to malnutrition and diseases like heart disease.

What amazes me is the vast number of relatively wealthy citizens that have bad teeth and do NOT spend the money for the necessary dental work.

Medical in rural areas of the nation may be few and far between due to the fact that most of our medical facilities require lots of dollars for equipment, administrative and medical personnel, and allowable medical facilities.

Urban areas are more likely to have the best hospitals and medical services due to the fact that there are more people per square mile that are paying for the facilities.

To learn more about this, see the *IRS webpage*: "Community Health Needs Assessment for Charitable Hospital Organizations - Section 501(r)(3)"

Segregated Cities and States:

Many of our states are split between the rural areas and the urban cities and suburbs and thus, may have both Republican and Democrat voters. This split caused tension and in some instances riots between the groups. For example; if the state has more Republican voters than Democrat, then the state turns

'Red' and not 'Blue'. Being a 'Red' state it then passes tax bills, transportation and other rules and laws that benefit the rural areas at the detriment of the urban areas. Many times this is reversed where the Blue outnumbers the Red and thus, the Red portions feel neglected.

Here is the problem. Let's say you need a new road and you live in the rural area where there are 10 houses per mile of the new road. The new road is estimated to cost $1,000,000 and thus $1,000,000 / 10 = $100,000 per household that will be using that road. There is virtually no way to tax or charge each household $100,000. But, if you are building a new road in the city where there are 1,000 houses per mile, then the cost comes down to $1,000 per household, which may be affordable, especially if charged over a 10-year period.

Now since both the rural and the urban group pay taxes, the rural group may get peeved at paying $1000 for a road in the city when they need a country road. The urban people may get annoyed at paying $1000 for a road in the rural area that they may never see or ride upon.

This dilemma has no solution and is multiplied by the numbers of roads needed; and by the numbers of other government or health care facilities needed. In short, we need to find a better tax system that is more equal for each of the 40-political groups.

Communications Deprived:
As just stated, there is a dollar problem between the amount of money collected in an area by a group and the amount spent in an area for a group. We still have groups of people in the U.S. that have no access to the Internet, they are too remote or two poor, or just not considered as needing it.

See "Massive Digital Divide for Native Americans is 'A Travesty'" REF: Mediashift.org

Transportation Deprived:

Americans love their vehicles and tend to hate public transportation even though it is safer, faster in many instances, and less expensive for most travel. See "Public Transportation Facts" REF: Apta.com

This author believes that the *Interstate Highway System* was the worst way to go and that a light rail system should have been the transportation system of choice. *Light rail* can use non-polluting electric energy instead of highly polluting fossil fuels. Light rail can be computer controlled to allow robot vehicles to self-guide from and to destinations. Light rail can be much faster with speeds up to several hundred miles per hour. Light rail only needs two-foot wide concrete or asphalt tracks and not trillions of tons of material for multilane highways that have to be rebuilt every 10 to 40 years. Light rail can be used for transport of very heavy commercial loads, and for passenger loads on the same tracks. Light rail can last from two to five times longer than fuel driven vehicles. Light rail can be designed for underground or on overhead lines thus, allowing for more space for buildings. Light rail can be built to go from building to building, even 30 stories in the air.

Our nation, along with many others will in about 50 years come to a halt as our fossil fuel supplies dwindle to a trickle, and our Interstate Highways fall apart due to lack of repair and replacement funding. We will be back to the horse and buggy, if we do not now start planning for and installing light rail or other environmentally and cost effective transportation systems.

See this author's manuals "Border Security Solved with High Speed Rail: Generating Jobs using Existing Solutions" and "High Speed Rail: What you Should Know!"

War:

War is a fool's paradise and we, the United States of America are the biggest fools on the planet. We have created a society of riches based on supplying the tools of war to virtually every nation on this earth. We continuously sell our 'old' weapons to people in other nations and then have to build 'new' weapons to protect ourselves from those that would use our 'old' weapons against us. We have since 1999 increased our yearly military budget from $199 billion to over $989 billion. {Estimated 2020 spending}.

The British Empire, the French, the Germans, the Chinese, the Japanese, the Romans all had great military armies and their excessive military expansion and spending eventually toppled each.

The U.S. Congress and government depend on its taxpayers to continue to supply the dollars to maintain the status quo. Nearly every state has some sort of military establishment be it a base or a weapons producer.

Pure Terror:

It is interesting that something as small as a virus can create pure terror in the overall population. The 911 incidents were caused by a handful of terrorists {19 total} that hijacked four aircraft. This one incident that at the time killed some 2,977 people started wars that killed tens of thousands of innocent people, men, women, and children. The wars ruined the lives of millions and after 17 years {March 2003 – March 2020} is still raging on and costing trillions.

The wars were created by a series of lies that managed to get people terrorized to the point they believed that the enemy was on their street and knocking at their doors in an effort to steal their children. Thousands of Americans took up arms and flew to foreign nations seeking revenge, only to return home in body bags or without limbs. Sad.

Nuclear Radiation:

I worked at a company called *Isotopes* for a year back in 1965.
Part of the job was spending weeks at the *Mercury Nevada
Atomic Energy* testing site. We were collecting air samples after
'shots', i.e., atomic bomb explosion tests. We did not wear
radiation badges at the time and the U.S. government in
subsequent years decided that those before 1965 and after 1965
were covered for radiation problems, but not those in 1965.
Yes, I as well as others did receive doses of radiation to the
point I was fired for setting off the radiation counters every
time I entered the sampling and analyzing room.

In the 1961 to 1962 period I was stationed in *Wakkanai Japan* and
we were downwind from the Russian testing site and on many
days were restricted to the barracks due to high readings of
nuclear radiation from a *Russian Atomic Bomb test.*

Over the years we have had in the United States, Japan, and in
Russia peacetime atomic energy electric generation plants self-
destruct. These events caused massive local disruption to
farms, towns, and people. We currently have nuclear power
plants located within a few miles of some of our largest cities
and population areas, and although nuclear plants are
considered very safe pollution free energy, there is always a
chance that one will experience a terrorist attack; or a natural
disaster like a hurricane, tornado, or earthquake that will
release massive doses of radiation into the atmosphere.

Our military uses *atomic weapons* and has thousands of nuclear
warheads, i.e. bombs, stored around the nation. Again, safety
is the major concern and most are very well tracked and safe
from harm, but … there is always that one in a million incident.

Our *nuclear power plants* and weapons occasionally need to be
deactivated and the waste products moved to a secure storage
site. This means that the material has to be moved by truck or

train through towns and cities and thus subject to accidents or terrorist hijackings.

Nuclear material is used in the medical and other industries and even a small amount in the hands of a terrorist can be used for making a 'dirty' bomb and scaring the ... out of thousands.

Cyber Warfare:

The average citizen has at least one computer or cell phone and he or she depends on this hand-held 'brain' 24/7. Take the machine away from them and they tend to go bonkers. Cyber warfare is the act of using programming or other means to block or distort the data that is provided. This today is equivalent to removing a person's brain from his or her body. Well almost, we absolutely are changing our lives to be dependent on instant data, self-diagnosing kitchen appliances, self-driving vehicles, automatic buying and purchasing of items, and much more.

A cyber attack can mess up our way of life in all sorts of ways that can be funny or tragic. Imagine someone taking over your car as you travel down the highway at 70 MPH or someone taking over your house door locks in the middle of the night. Yes, it can happen, for if it is computer generated or controlled it can be 'hacked' and controlled by those that wish you harm.

Chemical Attack on Food Sources:

In the past few years we have had contaminated lettuce, contaminated beef, and many other *food recalls* due to someone not properly handling our food. Much of the contamination is from humans that are excreting body fluids in our fields or not properly washing their hands after using toilet facilities. Some are due to improper cleaning of processing equipment or improper storage and storage temperatures.

Our food distribution systems allow for a basket of food to be packaged and distributed to many stores across many townships or states. This therefore can spread a food related disease to thousands within days or even hours.

To an enemy, this is a prime means of destroying a target, i.e. people, without destroying the target, i.e. the buildings and infrastructure. This is an ideal war, you get to eliminate the enemy but keep all the enemy's 'toys'.

There is also the possibility of a plant or meat source being contaminated by the *fertilizer* we used, or the *insecticide* used. Our food contains all sorts of elements that can cause problems due to our use of growth *hormones* and *Antibiotics*. Our major food suppliers are continuously experimenting with new chemicals and *DNA modifications* in an attempt to increase profits and supply. This can become very dangerous, very fast.

Animal sickness has in the past affected millions of chickens, fish, sheep, cows, etc. If one of these illnesses becomes transferable to humans, then we can sicken or kill thousands within days of providing the supply to our stores.

This too is a security problem in that some idiot or nation may figure out a way to add a poisonous contamination to our food supply in an attempt to wipe out an entire population.

Chapter # 06 – Robots Take Over:

If you watched one of the more popular space travel programs you may have seen the *Borg*, a race of half-human, half-robot beings that tended to work together as a unit. We have not yet gotten to the point of being the 'Borg', but we are fast approaching it. *Our cell phone generation* has come to the point that many people cannot shop, eat, sleep, work, or play without a cell phone in their ear. As the technology gets better that little hand-held computer/camera/phone will eventually become an implant in the brain or elsewhere on the body.

Robots were initially designed to take over jobs that were too tedious or dangerous for the average human. Each robot did one specific chore and was immobile. Today's robots may be equipped with *AI, Artificial Intelligence*, that allows the robot to think, make decisions, and then act on those decisions. Today's robots are more and more mobile so that each can travel during its job of cleaning your floor, delivering packages, disarming bombs, etc. Eventually, the robots will be building and maintaining other robots and thus, displacing man as the builder and master. Again, *Hollywood* has been on the forefront of this with its various movies and television shows that depict the future.

Automation:

People think of automation as being a machine that does a particular thing such as welding a seam on a car or installing a part onto another. These are examples of automation, but there are other not so obvious examples; such as *Google tm* being able to finish your 'search' pattern, or a sensor being able to stop your vehicle before it hits something. These 'thinking' automation items are replacing the human brain and reflex functions at an alarming rate.

It will eventually get to the point that human thought will no longer be needed, all the data, all the knowledge, all the actions will be controlled by automation and at the fingertips of all. I grew up reading a 21-volume set of encyclopedia cover to cover, several times. Today I have no need to read as I can speak to the television, computer, or little box on the counter and ask any question, getting a verbal answer in seconds.

Thus the question, does the earth need humans? Probably not as the robots require little to no food, much smaller housing, no need for schools or stores, and can replace themselves from a bucket of parts.

One Company Owns it All:

We at one time had a company for building, shipping, and then selling and maintaining just about each item we used. Today we have companies that can build, ship, and sell dozens to thousands of items fast and efficiently. We have *Amazon.com* that is fast becoming the only retail store in the world. As companies buy out their competition, to gain product, market share, and removal of competition we are approaching the fish in the pond scenario.

The fish in the pond are living together and eating the existing food, until the food becomes scarce. At that point the larger fish eat the smaller fish and this continues until there is only one gigantic fish left, and it starves to death due to no longer having any food.

We are approaching this point as the rich get richer and the poor get poorer, which will eventually lead to the poor either not being able to purchase the products that made the rich wealthy, or the poor dying off due to not being able to purchase the items that made the rich wealthy. When this happens, the rich will no longer have a sustainable market and thus will become the poor, but now not be able to supply the poor, and thus die.

Chapter # 07 – Nature Related

Nature is an integral part of our world and we need to not just respect it, but learn to live with it. Our *destruction* of the oceans, the waterways, the forest, and our grasslands is causing interference with the ways of nature that have been here before us, and will be here after us.

Animal Kingdom:

Most animals smartly are afraid of the human animal and avoid contact as much as possible. There are those that are not afraid due to physical size or having the capability to kill using self-contained poisons. These are elephants, lions, komodo dragons, snakes, frogs, fish, and some socialized mammals that hunt in groups.

We as humans are depleting their food sources, their water supplies, and the nesting areas for many of these creatures and each is adapting by losing their fear of humans and moving into our spaces. *Snakes in Florida*, monkeys in India, foxes and wolves in our suburbs, and rats in our cities are bringing new problems to our lives; including diseases that we normally would not be contacting. Many of the viral epidemics that we are facing originated from contact with Bats, birds, and other animals.

Insects:

The wonderful world of the insect is foreign to most of us, and many of us look upon an insect as an annoyance, rather than a benefit. Most spiders, bees, and other insects are beneficial to our lives as each helps to control other insects, pollinate plants, and provide foods or visual beauty. Some insects though have built-in protection in the form of poisons that can injure or kill. As we apply insecticides to the insect world they develop more and more resistance and the day may come when they due to their extreme numbers take over and use us for their food sources. Think it cannot happen, well think again as there are communities that have had to vacate due to killer bees and fire ants, and there are millions of buildings that have to be raised due to termites and other insect invasions.

Insects May Kill Us:

Common insects like the *House Fly* and *Mosquito* can carry diseases that can kill or at the minimum make us very sick. These are though only a few of the billions of insects that surround us humans. There are spiders, ants, lice, and some 925,000 others that can cause trouble for mankind. If these creatures ever evolve to the point of deliberately protecting themselves, and some already do, we may be in for a shock as tens of thousands of us die from bites, stings, and diseases carried.

Fish:

People love to look at fish swimming around in a fish tank and this is a big business. The problem is that when some persons get tired of their fish, they dump each into the rivers and ponds, thus introducing a 'foreign' species into the local waters. Chicago has the *Asian Carp*, a predator that must be kept out of Lake Michigan and the other Great Lakes. The fish was introduced to America over 30 years ago as a means of control algal build-up in sewage treatment plants, an experiment that has gone very wrong.

PS: Illinois and Chicago are spending billions of taxpayer dollars in an effort to bring this epidemic back under control.

Alligators:

Bensalem Pennsylvania a few years ago experienced alligators in their creeks. These are creeks where children go fishing and swimming, elder citizens walk the creek side paths, etc. It turns out that drug lords used young alligators and crocodiles as 'guard dogs' in their warehouse and drug manufacturing operations, and when the animals get too large to control, each is released into the wild. In this case the wild was the main city park.

Snakes:

Burmese Pythons were introduced into the Florida waterways several years ago and the snakes have no natural enemies, only man. Thus, animals like raccoons, and other snakes, fish, etc., have been reduced in population by up to 90% or more. The pythons are spreading into the Florida Keys, other parks and preserves, and are expected to spread throughout the southeast. Recently, there has been found that the snakes are mutating into other species, and that they also carry worms that are detrimental to other snakes and animals, thus killing each species. This is only one example of what can happen when a person that owns a foreign snake or creature releases it into the wide. PS: Florida is spending billions of taxpayer dollars in an effort to bring this epidemic back under control.

Bees:

If you live in the southwest you should be on the lookout for 'Killer Bees'. These once 'honey bees' were bred by scientists in Brazil as a hybrid between *African and European bees* in an effort to produce more honey. The experiment got out of control in 1957 and the *Africanized bees* are now in several nations. These bees produce less honey, are known to have up to 80,000 or more in a hive, and will attack the face, especially the eyes, of humans. Thus, they are very dangerous.

Bats and Birds:

These animals are essential to the environment as each feed on insects and other harmful items. The problem is that each also produces *'droppings'* and these droppings can cause human respiratory problems like pneumonia and contagious diseases.

Most backyard bird houses or feeders are not much of a threat, but a cave full of droppings, or an area of land being plowed up or bulldozed for new development may send up a cloud of dropping dust that can affect hundreds or thousands of people.

The 2011 movie *"Contagion"* is an example of what could happen, and in 2020 apparently did happen.

Monkeys:

The *Rhesus Macaque monkey* has taken over the streets of *New Delhi India* and these creatures are now robbing people, breaking into houses, and flat out terrorizing the citizens. There have even been reports of monkeys stealing human babies. The animals are seeking scarce food, and will attack men, women, and children in an effort to get something to eat.

There is a two-fold problem; the first being that humans are having too many babies and therefore, are taking over lands that were once habitat for the monkeys. Second is that the monkeys are not only having too many babies, but also that the Hindu religion protects life, including monkeys, and therefore the government has banned the killing of the monkeys.

Plants May Poison Us:

We all know of *Poison Ivy*, Poison Sumac, and Poison Oak, but did you know that *Oak Trees* can be poisonous to farm animals, and many fruits and nuts are poisonous to some people? There are all sorts of poisonous plants that we have learned to tolerate, such as some varieties of mushrooms, but if these plants ever evolve to the point of deliberately protecting themselves, and some already do, we may be in for a shock as tens of thousands of us die.

Natural but Unnatural:

These are things that do or can happen to the earth's population of which we have little control over, and that if large enough even may possibly eliminate life, as we know it.

Meteor Hit:

The earth has been hit by meteors many times in the past and if you visit the *Flagstaff Arizona* area you can visit *Meteor Crater* and see the massive size of the destruction caused by a relatively small meteor. A much larger meteor is thought to have hit off the coast of Central America that is thought to have killed the majority of prehistoric life on our planet.

We currently have few if any means of detecting potentially hazardous meteors or destroying one if it is coming our way.

We definitely do not have sufficient means of moving mass amounts of citizens out of the path of a meteor, or means of protecting them in the event of a direct or even indirect hit.

Moon Displacement:

We take the moon for granted, and we believe it will always be there in the sky for us to see. We depend on our moon for our tides and ocean health. If the moon for any of many reasons changes orbit or decides to come closer or leave all together, we are in trouble. The moon is our future, it will eventually be the base from which we launch adventures to other planets, mine for minerals, develop for medical purposes, or visit for recreational fun.

Comet Hit:

Similar to a meteor hit, a comet hit can be devastating and like a meteor comets are difficult to detect and destroy.

Sky on Fire:

Can the sky burn? For the most part, no even though it is composed of the elements that can burn, i.e. *Hydrogen* with the help of *Oxygen*. The sky is mostly *Nitrogen* and it is inert. The Hydrogen and other flammables like *Methane* are too diluted to catch on fire. That said, we might still be in trouble as the *Permafrost* melts and we release more flammables into the atmosphere each day.

Methane Gas:

As humans and animals digest food, each produces waste by-products and one of these is Methane Gas that can be flammable. As plants and garbage rot, each too produces Methane gas. As we drill for oil or mine coal some of the earth's methane gas is released into the atmosphere. As the *Permafrost of Alaska and Siberia* melts it too releases century old methane.

Methane is a *Greenhouse Gas* that produces *Carbon Dioxide* and Water, each also a heat trapping substance. It takes about 12 years for the methane in the atmosphere to change state, so during that time it can be burned if concentrated, which it usually is not. Many townships are now using captured, thus concentrated, methane gas from buried and rotting garbage dumps. Many farmers are capturing and using the methane from their bovines for generating heat and electricity for their farms.

The day may come when there is sufficient quantities of methane in our atmosphere that with the existing oxygen and a single spark, (example from a rocket heading out of earth's orbit) can set the sky on fire, which will consume much if not most of earth's oxygen, and thus we all die.

Ocean Burps Methane:

Deep in our oceans the temperature of the water is very cool, and in some areas it can be under pressure almost like solid ice. This coolness and pressure that covers rotted plant and animal life produces methane gas, and if the sea water warms, the gas rises and then can no longer be held in place. We therefore get what is considered a *Methane Burp*, a large volume of gas released suddenly. This Burp, therefore rises to the surface and can disrupt shipping, overturn some boats, and if there is a source of ignition, possibly catch on fire or explode.

Chapter # 08 – Opportunities Are Here:

The *'cause and effect'* of each of the items mentioned in this document can be either a bad thing or a good thing. Bad in that if nothing is done to solve or prevent the stated problems the human race can and probably will be in trouble. The good is that with thought and inspiration comes the solutions in the form of new procedures, new products, and thus new job opportunities.

Germ Contamination Suppressing Opportunities:

For example, we shop in all sorts of retail stores and we use our credit or debit cards in the countertop machines. These machines are only cleaned once or twice per day if at all and therefore, become a home for germs and other contaminants.

Gas station pumps are another place for passing germs to each other and these are never cleaned. These are just two of the thousands of items that are touched by your neighbors each day, and that can become 'homes' to deadly diseases. The opportunities for creating alternate 'no touch' or 'self-cleaning' areas are immense.

Pure Water Systems:

Pure water is a must for personal safety and for life in general and as our water supplies become more and more polluted, the need for less expensive water purification systems is booming. Not just in the United States but also in most nations. The need for processes and machinery that does not pollute provides opportunities for tens of thousands.

Heating and Cooling Systems:

We all use energy in the form of heating and cooling our buildings, in transportation, and in providing a means for manufacturing products we use. We have taken the baby steps to eliminate dirty coal and fossil fuels by generating power using natural solar and wind energy, but these too have certain harmful qualities to our resources and environments. The opportunity is to find more ways to produce the energy we need while simultaneously reducing harmful byproducts.

Clean Air Systems:

Currently clean air depends on filtering systems that may also contain electrostatic or *Ultraviolet* add-ons. These systems produce side effects, demand energy consumption, and require frequent maintenance. Inventions and devices that can decrease or eliminate air pollution should be and are a major source of ingenuity and job creation. Industrial plants that produce smoke and dangerous fumes need 'scrubbers' that remove the contaminations. The removed carbon and gasses can and should be used for creation of other materials and processes, and this opens up opportunities for inventors to come up with ideas and products that use industrial waste.

44

Gasoline and Oil Based Fuels:

Gasoline, jet fuels, and oils are being used at a rate that is not sustainable into the next decades and eventually will be depleted or so costly that only the super rich and the military will be able to afford each. The sun produces near infinite energy and thus should be the primary source of energy. It has been for thousands of years as the oil we use is basically stored sunlight energy. If you look to the past, about the 1960s, you will find that magazines like *'Popular Mechanics'* were full of ideas on how to harness the sun's energy. Again, this is an opportunity for jobs and profits.

Medical Opportunities:

The 2020 Covid-19 *Coronavirus Pandemic* showed the world how unprepared it was for a massive viral sickness. Not enough beds or isolation facilities, not enough testing kits, lack of protection for medical personnel, and the lack of an infrastructure that is 110% prepared for the sudden influx of tens of thousands of medical emergencies.

There will be many more *medical emergencies* and pandemics in our future and we need to be prepared for each with newer containment procedures, more isolation wards, better trained medical personnel, faster to the public test kits, and super fast methods of getting antiviral medications to those that need each.

The 2020 pandemic caused massive panic and economic hardships for tens of thousands of businesses and millions of citizens due to the amount of misinformation and the 'fear' factors. This opens up all sorts of opportunities for medical personnel, at-home care, disinfecting services, isolation facilities, test kit producers, laboratory services, secure ground transportation, and much more. Unfortunately, it also opened up religious last-rite services and burial services that are designed to keep viewers safe from the underlying diseases that caused the deceased.

Food Services:

A pandemic situation results in people being isolated to their homes and therefore needed nourishment and medical aid. *Grocery stores* are beginning to offer home delivery of food and some drug stores are offering free delivery of medications. This will require pickers, packers, drivers, and vehicle maintenance workers.

Companies like Amazon.com have increased temporary pay and are hiring 100,000 temporary workers in an effort to meet the new demand for product used to combat the Coronavirus.

Transportation Opportunities:

The automobile and the pickup truck are the two primary means of transportation in the world. Yes, there are bike cycles, motorcycles, trains, and aircraft available for getting from one place to another, and these are fine time-tested means of transport. Think about the cost and the pollution that is generated by fossil fueled vehicles. Not just the pollution for the burning of fossil fuels that produce a *pound of carbon* for every 25 miles driven, but also the cost and pollution from the millions of miles of road surface.

Concrete and asphalt are used in most areas to cover the dirt that was once used for 99% of our roads. These covering materials are costly and each does wear as millions of vehicles travel upon each. Additionally, there is the maintenance of line strip painting, and applying salt and other materials to keep the roads dry and safe. All of this could have been prevented back in 1958 and 1963 by not selecting an *Interstate Highway System*, but a national rail system instead. Rail has many advantages over roads in that rail transport can be faster, safer, programmable, energy efficient, and longer lasting.

There has been massive objection to *High-Speed Rail* in the United States. Other nations have had *HSR* for decades and move millions of passengers per year at reasonable cost and great safety, using electric energy generated from the sun and wind.

Population and Animal Control Opportunities:

We can design poisons, or raise hybrid predators that will kill off unwanted predators. Or we can come up with more advanced birth control that may slow the growth of various species that are overtaking the planet.

I am sure that someone, somewhere has an idea or two that will solve the population problems, short of war or creating a disease that wipes out half the world. We have over 6.5 billion people on earth, and many times that number of other creatures, and the earth may not be able to contain all of these numbers. We are seeing the beginning of the problem that will only get worse, unless we wake up and start applying smart solutions.

Touring and Tourist Opportunities:

This author has two other books that you should consider reading, the first is 'The Dead Blue & You', and the second is 'Environment Tours in the U.S.A.'. These books provide tons of opportunities for creating environmental tours.

The author's "Pioneers of Oceanography: Saga of the Robert D. Conrad" provides a look into the early {1960's} days of ocean science, and the book "Colony One...A Short Story" a quick look into what may be the future.

Index:

More About the Author:

The author has written several construction books for Craftsman Book Company and Prentice Hall. He has also designed and built homes. He has some 45-patented designs for Wet Process and Integrated Circuit Manufacturing equipment and was instrumental in the modification and setup of high-tech recording systems used to record the trips to Mars and our Moon. He designed the first 1-uv table top chart recorder and helped design the gas chromatograph used for analyzing the moon samples from the first moon landing.

Cover Picture:

In front of a city residence this wild coyote scratches itself in an

attempt to remove parasitic insects. The picture is of a Tucson Arizona desert landscape and contains a sign for 24-hour protection.

This picture represents the changes in our lives in that we now have to protect our property from people's greed, landscape with desert plants to save water, and be on a constant watch for wild animals that have been displaced from their natural habitat. The mailbox represents the beginning of the past where we cut down trees to make paper, and we now do most of our correspondence via cell phones and computer e-mail.

www.ingramcontent.com/pod-product-compliance
Lightning Source LLC
Chambersburg PA
CBHW030533220526
45463CB00007B/2813